I0469086

U.S. Fire Administration

The Critical Infrastructure Protection Process Job Aid

Emergency Management and Response-Information Sharing and Analysis Center

FA-313
2nd Edition: August 2007

FEMA

Table of Contents

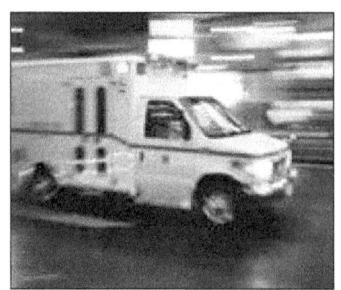

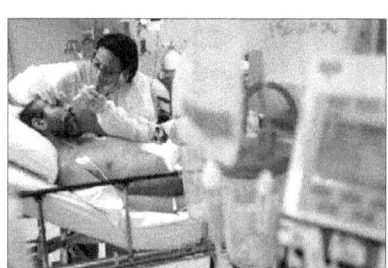

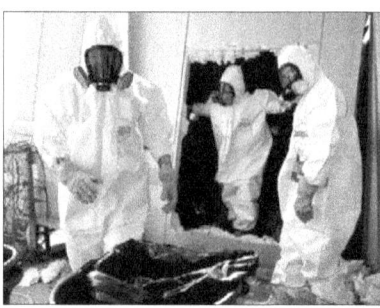

I. Introduction

A. Job Aid Purpose

- This Job Aid is a guide to assist leaders of the Emergency Services Sector (ESS) with the process of critical infrastructure protection (CIP).

- The document intends only to provide a model process or template for the systematic protection of critical infrastructures.

- It is not a CIP training manual or a complete road map of procedures to be strictly followed.

- The CIP process described in this document can be easily adapted to assist the infrastructure protection objectives of any community, service, department, agency, or organization.

B. Background

Homeland Security Presidential Directive – 7 (December 2003) established the requirement to protect national critical infrastructures against acts that would diminish the responsibility of federal, state, and local government to perform essential missions to ensure the health and safety of the general public. HSPD-7 identified the Emergency Services as a national critical infrastructure sector that must be protected from all hazards. The Emergency Management and Response—Information Sharing and Analysis Center (EMR-ISAC) activities support the critical infrastructure protection and resilience of Emergency Services Sector departments and agencies nationwide. The fire service, emergency medical services, law enforcement, emergency management, and 9-1-1 Call Centers are the major components of the Emergency Services Sector. These components include search and rescue, hazardous materials (HAZMAT) teams, special weapons and tactics teams (SWAT), bomb squads, and other emergency support functions.

1

C. EMR-ISAC Goals

- Promote awareness of the threats to and vulnerabilities of Emergency Services Sector (ESS) critical infrastructures via sector-wide information sharing.

- Encourage ESS prevention, protection, and resilience actions for all hazards, including man-made and natural disasters.

- Enhance survivability, continuity, and "response-ability" in all-hazards environments.

D. EMR-ISAC Major Tasks

- Collect critical infrastructure protection and resilience information having potential relevance for Emergency Services Sector (ESS) departments and agencies.

- Analyze all collected information to ascertain its importance and applicability to ESS organizations.

- Synthesize and disseminate emergent and consequential infrastructure protection and resilience information for the leaders, owners, and operators of the emergency services.

- Expedite distribution of Department of Homeland Security (DHS) all-source threat information to the validated ESS senior leaders.

- Develop instructional materials for CIP implementation and training needs.

- Provide professional assistance to ESS CIP practitioners via phone, e-mail, facsimile, and a web site.

E. EMR-ISAC Information Dissemination

- **CIP (FOUO) Notices.** Contain vital, actionable information published by the Department of Homeland Security (DHS) as needed regarding threats and vulnerabilities potentially affecting emergency plans and operations. (Forwarded **For Official Use Only** by those in vetted ESS leadership positions.)

- **CIP INFOGRAMs.** Contain four short articles issued weekly about protecting the critical infrastructures of emergency responders and their communities.

- **CIP Bulletins.** Contain timely homeland security information distributed as needed involving the infrastructure protection of the emergency services.

- **CIP Video.** DVD reviews the practice of CIP and the role of the EMR-ISAC.

- **CIP Job Aid.** Briefly explains how to practice CIP in a time-efficient and resource-restrained manner.

- **Homeland Security Advisory System Guide.** Provides recommendations for probable ESS actions before and during each level of the system.

F. Glossary

The definitions in this glossary are derived from language enacted in Federal laws and used in national plans, including the National Infrastructure Protection Plan, National Incident Management System, and National Response Plan.

- **All Hazards.** An approach for prevention, protection, preparedness, response, and recovery that addresses a full range of threats and hazards, including domestic terrorist attacks, natural and man-made disasters, accidental disruptions, and other emergencies.

- **Critical Infrastructure.** People, assets, systems, and networks, whether physical or virtual, so vital to the United States that their incapacity or destruction will have a debilitating impact on security, the nation's economy, public health or safety, or a combination of those matters.

- **Critical Infrastructure Protection (CIP).** CIP consists of the proactive activities to protect the indispensable people, physical assets, and communication/cyber systems from any degradation or destruction caused by all hazards.

- **Emergency Services Sector.** A system of preparedness, response, and recovery elements that forms the nation's first line of defense for preventing and mitigating the risk from man-made and natural disasters.

- **Emergency Support Function.** A grouping of government and certain private sector capabilities into an organizational structure to provide the support, resources, and programs needed to save lives, protect property, and restore essential services and critical infrastructure following domestic incidents.

- **Hazard.** Something that is potentially dangerous or harmful, often the root cause of an unwanted outcome.

- **Mitigation.** Activities designed to reduce or eliminate risks to persons or property or to lessen the actual or potential effects or consequences of an incident.

- **Preparedness.** The range of deliberate, critical tasks and activities necessary to build, sustain, and improve the operational capability to prevent, protect against, respond to, and recover from incidents.

- **Prevention.** Actions taken to avoid an incident or intervene to stop an incident from occurring.

- **Protection.** Actions to mitigate the overall risk to critical infrastructure people, assets, systems, networks, and functions, and their interconnecting links, from exposure, injury, destruction, incapacitation, or exploitation.

- **Risk.** The measure of potential harm from an all-hazards threat.

- **Resiliency.** The capability of people, assets, and systems to maintain functions during a disaster, and to expeditiously recover and reconstitute essential services after an all-hazards event.

- **Threat.** The intention and capability of an adversary (i.e., people and nature) to undertake actions that would be detrimental to critical infrastructures.

- **Vulnerability.** A weakness or flaw in an infrastructure that renders it susceptible to exploitation, disruption, damage, or incapacitation by all hazards.

II. CIP Overview

A. Premise

- Attacks on the personnel, physical assets, and communication/cyber systems of Emergency Services Sector (ESS) departments and agencies will weaken performance of mission essential tasks or prevent operations entirely.

- People (e.g., terrorists, criminals, delinquents, employees, hackers, etc.), nature (e.g., hurricanes, tornadoes, earthquakes, floods, wildfires, etc.); and hazardous materials (HAZMAT) accidents involving chemical, biological, radiological, or nuclear substances are the primary sources of attacks on critical infrastructures.

- Attacks on ESS personnel, physical assets, and communication/cyber systems are serious threats to the survivability, continuity, and response-ability of sector personnel and their operations.

- It is impossible to prevent all attacks (e.g., terrorism, natural disasters) against critical infrastructures.

- There will never be enough resources (i.e., dollars, personnel, time, and materials) to achieve complete protection of ESS critical infrastructures.

7

- There can be no tolerance for waste and misguided spending in the business of critical infrastructure protection (CIP).

B. Objectives

- To prevent or mitigate attacks on ESS critical infrastructures by people (e.g., terrorists, hackers, etc.), by nature (e.g., hurricanes, tornadoes, etc.), and by HAZMAT accidents.

- To protect the people, physical assets, and communication/cyber systems that are indispensably necessary for survivability, continuity of operations, response-ability, and mission success.

- To provide an analytical process to guide the systematic protection of ESS critical infrastructures by the application of a reliable decision sequence that assists sector leaders to determine exactly what needs protection and when security measures must occur.

- To provide a time-efficient and resource-restrained practice to ensure the protection of only those infrastructures that are critical for survivability, continuity of operations, response-ability, and mission success.

C. Additional Considerations

- From a municipal perspective, CIP is primarily about protecting those infrastructures unquestionably necessary for the continuity of crucial community services upon which citizen survivability depends.

- For Emergency Services Sector (ESS) departments and agencies, critical infrastructure protection (CIP) is foremost about protecting those internal infrastructures absolutely required for the survivability of emergency first responders and the preservation of their response-ability.

- CIP can be a tool to produce an American "mind-set" of protection awareness and confidence in our nation's security and prosperity. Given these new thoughts, it may evoke citizen behaviors that are fully supportive and cooperative with necessary protective measures.

- CIP may also be a means to change the behavior of terrorists. The proper protection of American critical infrastructures has the potential to develop a new "mind-set" among terrorists that their actions will be futile and not yield the results they seek.

D. CIP Process Review

- CIP involves the application of a systematic analytical process fully integrated into the plans and operations of ESS departments and agencies.

- It is a security related, time efficient, and resource-restrained practice intended to be repeatedly used by the leaders of the emergency services.

- The CIP process can make a difference only if applied by organizational leaders, and periodically reapplied when there have been changes in key personnel, physical assets, communication/cyber systems, or the general environment.

- The CIP process consists of the following five steps:

    1. *Identifying critical infrastructures* that must remain continuously intact and operational to accomplish ESS missions (e.g., fire suppression, emergency medical services, law enforcement, HAZMAT, search and rescue, emergency management, and 9-1-1).

    2. *Determining the threat* by all hazards against those critical infrastructures.

    3. *Analyzing the vulnerabilities* or weaknesses existing in the threatened critical infrastructures.

4. *Assessing risk* of the degradation or loss of credibly threatened and vulnerable critical infrastructures.

5. *Applying protective or resiliency measures* where risk is unacceptable to prevent the threat, protect the credibly threatened and vulnerable critical infrastructures, or ensure the rapid restoration of critical infrastructures after an all-hazards attack.

III. CIP Process Methodology

A. Identifying Critical Infrastructures

1. Identifying critical infrastructures is the first step of the CIP process.

2. The remaining steps of the CIP process cannot be initiated without the accurate identification of an organization's critical components.

3. ESS critical infrastructures are those personnel, physical assets, and communication/cyber systems that are indispensably essential for the survivability, continuity of operations, response-ability, and mission success of ESS departments and agencies.

4. Critical infrastructures are the people, things, or systems that will seriously degrade or prevent survivability, continuity, response-ability, and mission success if they are not continuously intact and operational.

5. The following are some examples of potential ESS critical infrastructures:

   a. Firefighters, police, paramedics, and emergency medical technicians.

   b. Fire, police, and emergency medical vehicles, equipment, stations, and communications systems.

   c. Computer-aided dispatch and computer networks.

   d. 9-1-1 Centers (Public Safety Answering Points).

   e. Water pumping stations and distribution systems.

   f. Major roads and highways.

   g. Key bridges and tunnels.

   h. Local and regional medical facilities.

6. Despite many similarities, the differences in personnel, physical assets, and communication/cyber systems among individual ESS departments and agencies necessitate that senior leaders identify their own critical infrastructures.

7. Remember that protection measures cannot be implemented if what needs protection is unknown!

8. Following the disasters in New York City (September 2001) and New Orleans (August 2005), ESS organizations in those two cities continued to serve their citizens. However, their ability to do so was tremendously degraded for a period of time given the unprecedented losses of first responder critical infrastructures.

B. Determining the Threat

1. Determining the threat from all hazards (e.g., terrorists, criminals, hurricanes, tornadoes, earthquakes, floods, wildfires, etc.) against identified critical infrastructures is the second step of the CIP process.

2. A threat is the potential for an attack from people or nature or a combination of these. It also includes the probability of a HAZMAT accident.

3. The remaining steps of the CIP process depend upon whether or not an organization's critical infrastructures are threatened.

4. Three examples of credible threats against ESS critical infrastructures:

    a. National intelligence sources warn that suspected terrorists may attempt to drive a vehicle-borne improvised explosive device into the police headquarters of a large city. (This example demonstrates a terrorist threat against all components of a police department's critical infrastructures.)

    b. Police cite increasing incidents of individuals breaking into fire departments to steal valuable equipment and damage apparatus. (This example demonstrates a criminal threat against the physical assets of a fire department's critical infrastructures.)

    c. The National Weather Service forecasted that the eastern and southern coasts of the United States will experience eight major hurricanes between 1 June and 30 November. (This example demonstrates a nature threat against all critical infrastructure components of all emergency departments and agencies in a designated area.)

5. A determination of credible threat must be made for each critical infrastructure identified in step one.

6. If there is no threat of an attack against one of an organization's critical infrastructures, then the CIP process can stop here for that particular asset.

7. When there is a credible threat of an attack against a department's critical infrastructures, prior to proceeding to the next step of the CIP process, it is necessary to determine exactly which critical infrastructures are threatened and by whom or what is each of these infrastructures threatened.

8. Leaders should concentrate only on those threats that will dangerously degrade or prevent survivability and response-ability.

9. Resources should be applied to protect only those critical infrastructures against which a credible threat exists!

C. Analyzing the Vulnerabilities

1. Analyzing the vulnerabilities of only credibly threatened infrastructures is the third step of the CIP process.

2. This step requires an examination of the security vulnerabilities (i.e., weaknesses rendering the infrastructures susceptible to degradation or destruction) in each of the threatened infrastructures determined in step 2 of the CIP process.

3. There are two types of vulnerabilities to consider in the CIP process:

    a. A weakness in a critical infrastructure that makes the infrastructure susceptible to disruption or loss from an attack by human adversaries.

    b. A weakness in a critical infrastructure that will further weaken or completely deteriorate as a result of a natural disaster or HAZMAT accident.

4. An efficient vulnerability analysis will examine only the credibly threatened infrastructures from the "threat point of view." This is particularly useful when trying to get into the minds of terrorists or criminals to understand how they think and, therefore, how they will behave.

5. The analysis will seek to comprehend the ways by which threats from adversaries, nature, or HAZMAT accidents might disrupt or destroy the examined infrastructure because of existing vulnerabilities.

6. The following are three examples of vulnerabilities:

   a. An ESS facility is vulnerable if located in a tornado-prone area, but has not been engineered to withstand a tornado.

   b. Emergency services personnel and equipment are vulnerable to injury, death, and destruction if not properly trained to protect against secondary explosions at incidents of terrorism.

   c. Emergency services personnel are vulnerable to acute illness or death if not vaccinated for hepatitis, tetanus, influenza, etc.

7. If a threatened infrastructure has no vulnerabilities, then the CIP process can stop here for that particular asset.

8. The CIP process should proceed to the fourth step only for those threatened infrastructures having vulnerabilities.

9. The protection of threatened and vulnerable infrastructures and the mitigation or elimination of existing weaknesses cannot be accomplished without knowing what or where the vulnerabilities are!

D. Assessing Risk

1. Assessing risk of the degradation or loss of a critical infrastructure is the fourth step of the CIP process.

2. Threatened and vulnerable critical infrastructures from step 3 of the CIP process are a high priority for assessing the risk of degradation or incapacitation.

3. Focusing on each high priority infrastructure, decision makers must evaluate the cost of protective or resiliency measures in terms of available resources (e.g., personnel, time, money, and materials).

4. The determined costs of protective or resiliency measures (i.e., doing something) for each high priority infrastructure are now weighed against the results of the degradation or loss of that infrastructure (i.e., doing nothing and accepting risk).

5. Risk is unacceptable if the impact of the degradation or loss of an infrastructure (i.e., from doing nothing) will be potentially catastrophic.

6. Three examples of risk assessment follow:

    a. ESS leaders assess that the incapacitation of their facility in a tornado-prone area would seriously disrupt continuity of operations and response-ability and, therefore, decide that the risk is unacceptable.

    b. ESS chief officers assess that the risk of injury or death from secondary explosions at incidents of terrorism is too great, and that doing nothing about this risk is unacceptable.

    c. The emergency medical services (EMS) leadership assess that there is significant risk to EMS personnel for acute illness or death from hepatitis, tetanus, influenza, etc. They determine that something must be done to protect employees against known pathogens.

7. If the impact of the degradation or loss of an infrastructure is not considered remarkable, then decision makers can temporarily decide to accept risk and do nothing until resources become available.

8. For the infrastructures that are risk adverse and require protection, the CIP process must proceed to the final step for the immediate application of protective or resiliency measures.

9. Failure to assess risk can result in the inefficient application of resources and a subsequent reduction in operational effectiveness.

E. Applying Protective or Resiliency Measures

1. Applying protective or resiliency measures to credibly threatened and vulnerable critical infrastructures unacceptable to risk is the fifth and last step of the CIP process.

2. Protective or resiliency measures are any actions that prevent the threat, protect the credibly threatened and vulnerable critical infrastructure, or ensure the swift restoration of critical infrastructures after an all-hazards attack.

3. Protective or resiliency measures are applied to high priority, risk-adverse infrastructures that necessitate the allocation of resources to preserve the ability of emergency first responders to efficiently perform their services.

4. Community leaders should decide the order in which credibly threatened and vulnerable critical infrastructures that are unacceptable to risk will receive the allocation of resources and application of protective or resiliency measures.

5. Possible protective and resiliency measures differ in terms of feasibility, expense, and effectiveness. Additionally, they can be simple or complex actions limited only by imagination and creativity.

6. The following are three examples of protective or resiliency measures:

   a. ESS leaders decide that the incapacitation of their facility in a tornado-prone area will unacceptably disrupt continuity of operations and response-ability and, therefore, agree to promptly acquire the resources to alter the structure to withstand a tornado.

   b. ESS chief officers resolve that the risk of injury or death from secondary explosions at incidents of terrorism is too great. Consequently, they arrange for the proper personnel training to prevent or protect against injury or death from secondary explosions.

c. The emergency medical services (EMS) leadership determine that there is significant risk to EMS personnel for acute illness or death from hepatitis, tetanus, influenza, etc. As a protective measure they ensure that every employee has been vaccinated against known pathogens.

7. In few instances, there may be no effective means to protect a critical infrastructure. Sometimes, prohibitive costs or other factors make the application of protective and resiliency measures impossible.

8. Decisions requiring the application of protective and resiliency measures will influence personnel, time, and material resources, and drive the security budget.

9. High priority, risk-adverse infrastructures should be considered a loss to plans and operations if protective or resiliency measures have not been implemented.

IV. CIP Process Question Navigator

DIRECTIONS: As a time-efficient tactic when reviewing or reapplying the CIP process, answer the following questions for each infrastructure of your department or agency. Alternatively, use the decision matrix seen at the next page.

- Is the person, thing, or system part of the organization's infrastructure?

    If NO, stop. If YES, ask:

- Is this infrastructure essential for survivability, continuity, and response-ability?

    If NO, stop. If YES, ask:

- Is there potential for an attack from people or nature against this critical infrastructure, including the probability for a HAZMAT accident?

    If NO, stop. If YES, ask:

- Is the threat of an attack against this critical infrastructure a truly credible one?

    If NO, stop. If YES, ask:

- Is there a security vulnerability (or weakness) in the threatened critical infrastructure?

    If NO, stop. If YES, ask:

- Does this vulnerability (or weakness) render the critical infrastructure susceptible to disruption or loss?

    If NO, stop. If YES, ask:

- Is it acceptable to assume risk and delay the allocation of resources and the application of protective or resiliency measures?

    If YES, stop. If NO, then:

- Apply protective or resiliency measures to protect this critical infrastructure as soon as available resources permit.

V. Infrastructure Protection Decision Matrix

DIRECTIONS: Complete the matrix for each infrastructure of your department or agency.

Is the infrastructure a critical one?

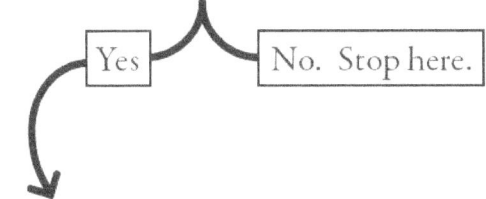

Is the critical infrastructure threatened?

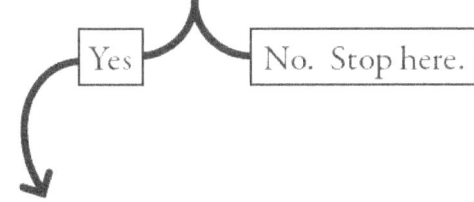

Is the critical infrastructure vulnerable?

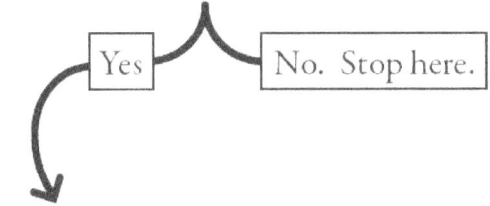

Is risk of degradation or loss acceptable?

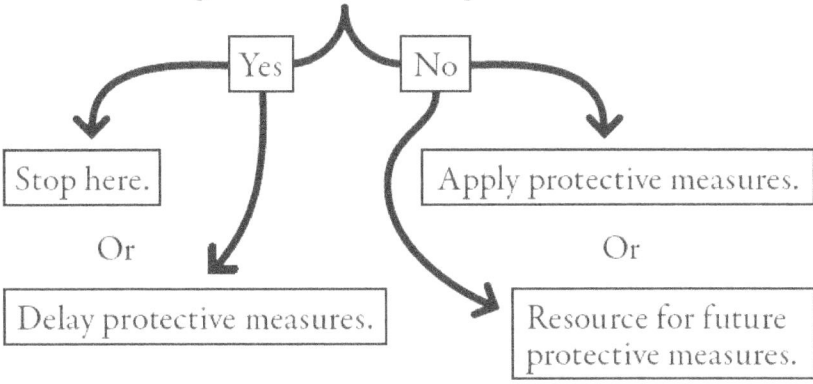

VI. Establishing a CIP Program

A. Justification

- A quality CIP program supports the protection or resiliency of the people, physical assets, and communication/cyber systems upon which survivability, continuity of operations, response-ability, and mission accomplishment depend.

- The threat of domestic and transnational terrorism should provide ESS leaders with sufficient justification to implement a CIP program within their organizations.

- If the threat of terrorism itself does not motivate action, then remember that the CIP process can also mitigate or eliminate the degradation or destruction of ESS departments and agencies caused by nature and HAZMAT accidents.

- The CIP program should be integrated as a component part of an organization's security and emergency preparedness plans, as well as the conduct of operations.

B. Program Manager

- Critical infrastructure protection is primarily leader business. The emergency department or agency chief, commander, or director appoints a program manager from among the senior leadership of the organization.

- The program manager performs the following functions:

    1. Administer the CIP program and maintains its value, relevance, and currency.

    2. Prepare, obtain approval for, and publish the program's purpose, strategic goals, and immediate objectives.

    3. Lead the effort to acquire the resources necessary to implement protective or resiliency measures.

4. Initiate actions that provide protection or resiliency for the organization's critical infrastructures against all hazards.

C. Program Development and Management

- The emergency department or agency chief, commander, or director institutes the organization's CIP program and delegates authority to a manager.

- The following program development and management steps are recommended:

    1. Select the program manager from among the senior decision makers of the organization.

    2. Firmly establish the relationship between the organization's mission and the purpose for critical infrastructure protection.

    3. Acquire the support of the department senior and junior leadership, and orient the CIP program primarily to them.

    4. Focus the program on the practice of the CIP process contained in this guide.

    5. After determining which critical infrastructures are risk adverse (i.e., the outcome of step 4 of the CIP process), aggressively seek the resources required to apply protective or resiliency measures as soon as possible.

    6. Revise the department security and emergency preparedness plans to include the CIP program and the critical infrastructures that demand protective or resiliency measures.

    7. Brief all department personnel regarding the revised plans and ensure awareness of actions they can take to support applied protective or resiliency measures.

8. Practice operations security (i.e., protecting sensitive information) concurrently with CIP.

9. Remain vigilant for threat advisories and new CIP trends, methods, and conditions disseminated by the EMR-ISAC.

10. Maintain the program by reapplying the CIP process when there have been changes in key personnel, physical assets, communication/cyber systems, and the general environment; however, attempt to do so at least semi-annually.

D. Contact Information

The Emergency Management and Response—Information Sharing and Analysis Center (EMR-ISAC) will provide CIP consultation or assistance (via telephone, electronic mail, or facsimile) to any emergency organization practicing the CIP process or establishing a CIP program. Contact the EMR-ISAC at 301-447-1325, or by electronic mail at emr-isac@dhs.gov. Interested personnel can also visit the EMR-ISAC web site at http://www.usfa.dhs.gov/emr-isac. This information is current as of August 2007.